W9-CPY-262

Wild About Wheels

FIRE TRUCKS

by Nancy Dickmann

PEBBLE
a capstone imprint

Pebble Emerge is published by Pebble, an imprint of Capstone.
1710 Roe Crest Drive
North Mankato, Minnesota 56003
www.capstonepub.com

Copyright © 2022 by Capstone. All rights reserved. No part of this publication may be reproduced in whole or in part, or stored in a retrieval system, or transmitted in any form or by any means, electronic, mechanical, photocopying, recording, or otherwise, without written permission of the publisher.

Library of Congress Cataloging-in-Publication Data
Names: Dickmann, Nancy, author.
Title: Fire trucks / by Nancy Dickmann.
Description: North Mankato, Minnesota : Pebble, [2022] | Series: Wild about wheels | Includes bibliographical references and index. | Audience: Ages 6–8 | Audience: Grades 2–3 | Summary: "Smoke pours out of a window. Call 911. A fire truck is on its way! Fire trucks carry the equipment needed to put out fires and help people in emergencies. Young readers will find out about fire trucks, their main parts, and how these important vehicles are used"—Provided by publisher.
Identifiers: LCCN 2020025536 (print) | LCCN 2020025537 (ebook) |
  ISBN 9781977132345 (hardcover) | ISBN 9781977133281 (paperback) |
  ISBN 9781977154231 (ebook pdf)
Subjects: LCSH: Fire engines—Juvenile literature.
Classification: LCC TH9372 .D53 2022 (print) | LCC TH9372 (ebook) | DDC 628.9/25—dc23
LC record available at https://lccn.loc.gov/2020025536
LC ebook record available at https://lccn.loc.gov/2020025537

Image Credits
Alamy: Dembinsky Photo Associates/David Traiforos, 5; Capstone Studio: Karon Dubke, 21 (art supplies); Getty Images: Monkey Business Images, 9; iStockphoto: Alena Paulus, 8, ollo, 11, simonkr, 13; Newscom: Kyodo, 14; Shutterstock: Dariush M, 17, Giovanni Love, 16, Jeff Thrower, 6, Keith Muratori, 15, Mark Taylor Cunningham, 7, mikeledray, cover, back cover, Red Orange, 21 (drawing), Rob Wilson, 18¬–19, stockphotograf (background), throughout, Ted Pendergast, 12, Toa55, 10

Editorial Credits
Editor: Amy McDonald Maranville; Designer: Cynthia Della-Rovere; Media Researcher: Eric Gohl; Production Specialist: Katy LaVigne

All internet sites appearing in back matter were available and accurate when this book was sent to press.

Table of Contents

Words in **bold** are in the glossary.

WHAT FIRE TRUCKS DO

A building is on fire! Flames lick at the roof. Smoke pours out of the windows. People are trying to get to safety.

Here comes a fire truck! Its **sirens** wail. The lights flash. Firefighters leap out to tackle the blaze. The truck has all the tools they need. They will put out the fire.

Fire trucks carry tools for other jobs too. Fire trucks have tools for rescuing people. Some tools are used to cut people out of crashed cars.

If a building collapses, people may be trapped inside. The truck has tools like axes and **fire hooks** to help. Other fire trucks carry boats for water rescues.

Look Inside

A firefighter drives the fire truck. They can sit in the **cab**. There are many controls here. The fire captain sits in front too. They can use the radio.

There are more seats behind the cab. Up to six firefighters can sit there. **Air packs** are stored behind the seats. Air packs help firefighters breathe when the air is smoky.

Most fire trucks carry water. They have a big tank in the back. Firefighters spray this water on fires. The truck has a **pump**. It sucks up water. It pushes it through hoses.

A fire truck carries many different hoses. They are stored inside. When the truck arrives at a fire, the firefighters unpack the hoses. Firefighters attach them to the pump.

A fire truck has panels on each side. Firefighters open them to find the tools they need. Some tools pull down walls. Wrenches open **fire hydrants**. Fans suck out smoke. Special cameras show the firefighters where the fire is hottest.

Look Outside

Most fire trucks have ladders. Some trucks have really long ones! An **aerial ladder** has sections that slide out. It can reach high and swivel around.

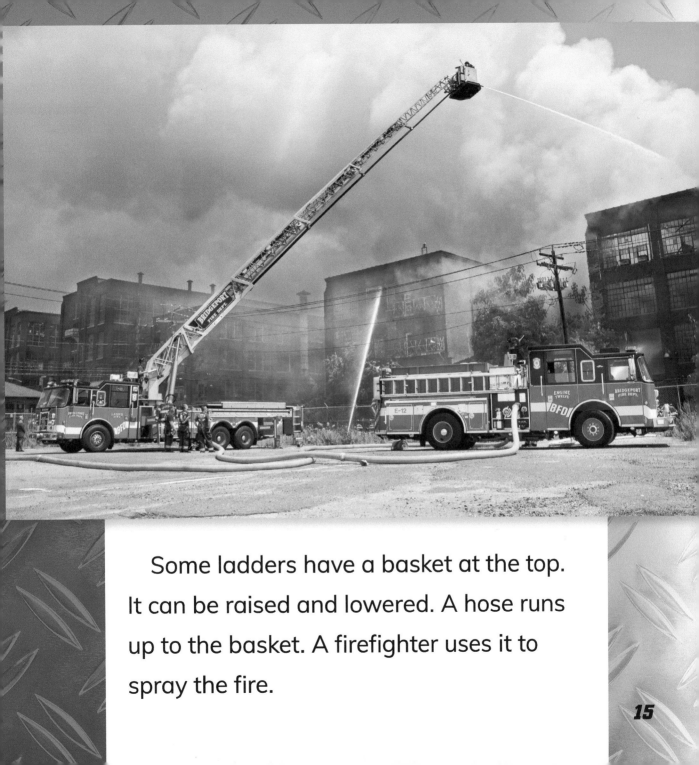

Some ladders have a basket at the top. It can be raised and lowered. A hose runs up to the basket. A firefighter uses it to spray the fire.

Many fire trucks have a water cannon. It is mounted on the top. A firefighter can aim it in any direction.

The cannon shoots water at the fire. The water can travel a long way. Sometimes the water is mixed with **foam**. Foam helps put out fires even faster.

FIRE TRUCK DIAGRAM

ladder

lights

cab

basket

tool panels

DESIGN A FIRE TRUCK

There are many types of fire trucks. Some fight wildfires. Others work at airports. Some are not trucks at all! Fireboats patrol rivers and ports.

Can you design a fire truck? How will it be used? What kinds of fires will it fight? What tools does it need? Draw and label a picture of your fire truck.

Glossary

aerial ladder (AIR-ee-ul LAD-ur)—a large ladder mounted on a fire truck that can extend and swivel around

air pack (AIR PAK)—the mask and tank that a firefighter wears in order to breathe in smoky situations

cab (KAB)—the compartment at the front of a vehicle, where the driver sits

fire hook (FYRE HUK)—a pole with a hook used to tear down walls

fire hydrant (FYRE HIGH-drunt)—a pipe in the street that provides water for firefighters

foam (FOME)—a substance made from tiny bubbles; some types of foam help put out fires

pump (PUMP)—a tool that sucks up water and pushes it out, such as through a hose

siren (SIGH-run)—a tool that makes a loud warning sound

Read More

Dittmer, Lori. *Fire Trucks*. Mankato, MN: Creative Education, 2019.

Keppeler, Jim and Jill. *Fire Trucks*. New York: PowerKids Press, 2020.

Internet Sites

A Day in the Life: Firefighter
www.youtube.com/watch?v=gbL45xX6p6E

Twenty Surprising Things about Fire Trucks Most People Need to Know
www.hotcars.com/20-surprising-things-about-fire-trucks-most-people-need-to-know/

Wonderopolis: Why Are Fire Trucks Red?
www.wonderopolis.org/wonder/why-are-fire-trucks-red

Index